Focus On ELEMENTARY Geology

Laboratory Notebook
3rd Edition

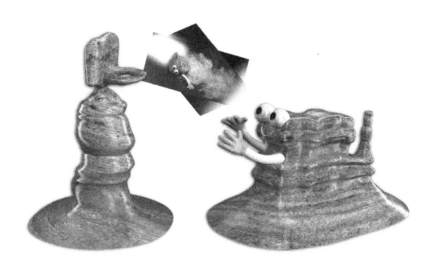

Rebecca W. Keller, PhD

Real Science-4-Kids

Illustrations: Janet Moneymaker

Copyright © 2019 Gravitas Publications Inc.

All rights reserved. No part of this publication may be reproduced, stored in a retrieval system, or transmitted, in any form or by any means, electronic, mechanical, photocopying, recording, or otherwise, without prior written permission from the publisher. No part of this book may be reproduced in any manner whatsoever without written permission.

Focus On Elementary Geology Laboratory Notebook—3rd Edition
ISBN 978-1-941181-40-9

Published by Gravitas Publications Inc.
www.gravitaspublications.com
www.realscience4kids.com

A Note From the Author

Hi!

In this curriculum you are going to learn the first step of the scientific method:

Making good observations!

In the science of geology, making good observations is very important.

Each experiment in this notebook has several different sections. In the section called *Observe It*, you will be asked to make observations. In the *Think About It* section you will answer questions. There is a section called *What Did You Discover?* where you will write down or draw what you observed from the experiment. And finally, in the section *Why?* you will learn about the reasons why you may have observed certain things during your experiment.

These experiments will help you learn the first step of the scientific method and......they're lots of fun!

Enjoy!

Rebecca W. Keller, PhD

Contents

Experiment 1	GEOLOGY EVERY DAY	1
Experiment 2	SMASHING HAMMERS	9
Experiment 3	MUD PIES	26
Experiment 4	THE SHAPE OF EARTH	36
Experiment 5	MUD VOLCANOES	44
Experiment 6	ALL THE PARTS	53
Experiment 7	EDIBLE EARTH PARFAIT	64
Experiment 8	WHAT'S THE WEATHER?	70
Experiment 9	HOW FAST IS WATER?	82
Experiment 10	WHAT DO YOU SEE?	90
Experiment 11	MOVING IRON	100
Experiment 12	WHAT DO YOU NEED?	108

Experiment 1

Geology Every Day

Even before geology became a science, people observed rocks and landscapes, mountains, lakes, and rivers. In this experiment you will make observations about how geology affects your daily life.

I. Think About It

Think about where you live.

❶ Do you live in a city or in the country?

_____

❷ Is your house built with concrete, stone, brick, or some combination of these?

_____

❸ Where did the builders find the concrete, stone, or brick for your house?

_____

_____

_____

❹ Do you live near mountains or on the plains? Near a lake, the ocean, or in the desert? Describe what you think the area where you live is like.

_____

_____

_____

_____

❺ What is the weather like where you live? Do you have lots of rain, tornadoes, or hurricanes? Describe some kinds of weather that happen where you live.

_____
_____
_____
_____
_____

❻ Do you live near a volcano? Are there earthquakes where you live? What do you think it is like to live in these areas?

_____
_____
_____
_____

❼ How much does the geology of where you live affect your life? In what ways is your life affected?

_____
_____
_____
_____
_____
_____

II. Observe It

❶ Make a list of all the geological features you see in a day, such as mountains, rivers, lakes, or an ocean.

_____
_____
_____
_____
_____
_____
_____

❷ Observe the weather where you live and write down what you experience—hot weather, cold weather, rain, fog, snow, drought, etc.

_____
_____
_____
_____
_____
_____
_____

❸ List the kinds of animals that live near you. Write down the names of any wildlife that you see, such as foxes, coyotes, squirrels, snakes, lizards, ants, bears, sparrows, or any other kind of wildlife.

_____
_____
_____
_____
_____
_____
_____
_____

❹ List the types of plants that live near you. Write down the plants that grow around your home, such as green grass, flowers, cactus, or weeds.

_____
_____
_____
_____
_____
_____
_____

III. What Did You Discover?

❶ Do the mountains, rivers, oceans, or plains affect the kinds of wildlife you see? Why or why not?

_____

_____

_____

❷ Does the weather affect the mountains, rivers, or oceans? Why or why not?

_____

_____

_____

❸ Does the kind of soil you live near affect what kinds of plants can grow? Why or why not?

_____

_____

_____

❹ Does the area where you live affect the kind of house you live in? Why or why not?

_____

_____

_____

IV. Why?

If you live near a volcano, or if the place where you live experiences earthquakes, you might be very aware of the geology around you. However, it is easy to forget that geology affects our daily lives even without earthquakes and volcanoes. Where your house is located, the kind of plants and animals you see, the weather you experience, and the view from your house are all determined by geology.

V. Just For Fun

Imagine what it might be like to live on the Moon. What would the landscape look like—flat, mountainous, or...? Where would you want your house to be located? What would it be made of? Would you have any weather? Would you see animals and plants, and if so, what kinds?

On the next page, draw a picture of your home on the Moon. If you would like to write down some of your ideas first, you can use the lines below.

_____

_____

_____

_____

_____

Home on the Moon

Experiment 2

Smashing Hammers

Introduction

In this experiment you will explore using a simple tool that geologists use for many purposes—the hammer.

I. Think About It

❶ What do you think will happen if you smash a banana with a plastic hammer?

_____

_____

❷ What do you think will happen if you smash a hardboiled egg with a metal hammer?

_____

_____

❸ Which do you think would work better for smashing a potato, a plastic hammer or a metal hammer? Why?

_____

_____

❹ What is inside a hardboiled egg? Is it hard or soft?

_____

❺ What is inside a rock? Is it hard or soft?

_____

❻ Is a banana soft or hard?

_____

II. Observe It

❶ Take the plastic hammer and smash a piece of banana. In the box below, write or draw what happens.

❷ Take the metal hammer and smash another piece of banana. In the box below, write or draw what happens.

❸ Take the plastic hammer and smash a hardboiled egg. In the box below, write or draw what happens.

Experiment 2: Smashing Hammers

❹ Take the metal hammer and smash another hardboiled egg. In the box below, write or draw what happens.

❺ Take the plastic hammer and smash a potato. In the box below, write or draw what happens.

❻ Take the metal hammer and smash another potato. In the box below, write or draw what happens.

❼ Take the plastic hammer and smash a rock. In the box below, write or draw what happens.

❽ Take the metal hammer and smash another rock. In the box below, write or draw what happens.

Experiment 2: Smashing Hammers 19

III. What Did You Discover?

❶ What happened when you used the plastic hammer to smash a banana?

_____

_____

_____

❷ What happened when you used the metal hammer to smash a banana?

_____

_____

_____

❸ What did you discover when you used the plastic hammer to smash a hardboiled egg?

_____

_____

_____

❹ What did you discover when you used the metal hammer to smash a hardboiled egg?

_____

_____

_____

❺ What did you discover when you used the plastic hammer to smash a potato?

___

___

___

❻ What did you discover when you used the metal hammer to smash a potato?

___

___

___

❼ What did you discover when you used the plastic hammer to smash a rock?

___

___

___

❽ What did you discover when you used the metal hammer to smash a rock?

___

___

___

IV. Why?

Geologists use different tools to study rocks, minerals, and Earth's layers. When a geologist needs to break open a rock to see what's inside, a rock hammer is the most commonly used tool. Smaller rock hammers that are lightweight can help geologists break open small rocks while preserving delicate gems or other structures. Larger rock hammers could destroy these small features.

Larger, heavier rock hammers are good for breaking open bigger rocks to expose fossils or larger gem structures. Smaller rock hammers might not be heavy enough to break open these bigger rocks.

In this experiment you used two different types of hammers—a lightweight plastic hammer and a heavier metal hammer. You probably discovered that the lightweight plastic hammer was able to smash the softer items and the heavier metal hammer could smash both the softer and the harder items.

A geologist will choose a lightweight hammer to break open smaller or softer rocks and a heavyweight hammer to break open bigger and harder rocks. Practice makes it easier for a geologist to choose the right hammer to use with a particular rock sample.

V. Just For Fun

Use a magnifying glass or hand lens to examine the objects you smashed in this experiment. Use the following boxes to write or draw what you observe.

BANANA

Unsmashed

Smashed With Plastic Hammer

Smashed With Metal Hammer

HARDBOILED EGG

Unsmashed

Smashed With Plastic Hammer

Smashed With Metal Hammer

POTATO

Unsmashed

Smashed With Plastic Hammer

Smashed With Metal Hammer

ROCK

Unsmashed

Smashed With Plastic Hammer

Smashed With Metal Hammer

Experiment 3

Mud Pies

I. Observe It

❶ Go outside to your backyard, to a park, or any place where there is dirt that can be collected. Using a small shovel, dig a sample of the dirt and put it into a small pail or plastic container.

❷ Look through the dirt sample you collected. In the space below, draw what you observe.

❸ Using your hands, separate the dirt from any small or large rocks you collected in your sample.

❹ Notice that rocks are different from dirt. Explain, write, or draw how you can tell the difference between rocks and dirt.

_____
_____
_____
_____
_____
_____

❺ Take about .25 liter (1 cup) of the dirt mixture and put it in a tall, clear glass container. Pour water into the container so that it covers the dirt completely. Make sure that there is 5-8 centimeters (2-3 inches) of water above the dirt.

❻ Stir the dirt gently with your fingers. Watch what happens to the water that was above the dirt and observe what happens to the rocks. Write or draw your observations below.

_____
_____
_____
_____
_____

❼ **Allow the mixture to settle. Observe what happens and write or draw your observations below.**

_____
_____
_____
_____
_____
_____

II. Think About It

❶ Can you tell the difference between rocks and soil (dirt)? If so, can you explain this difference?

_____
_____
_____
_____
_____
_____

❷ When you added water to the soil, mixed it up, and allowed it settle, what did you observe?

_____
_____
_____
_____
_____

❸ Add 60 milliliters (1/4 cup) of flour to the mixture. Stir the mixture and allow it to settle. Observe what happens and write or draw your observations on the following page.

III. What Did You Discover?

❶ What are some of the differences you observed between rocks and dirt?

_____

_____

_____

❷ When you added water to the rock/dirt mixture, did the water get cloudy? If so, why do think that happened?

_____

_____

_____

❸ When you allowed the rock/dirt mixture to settle, did you observe layers forming? If so, why do you think this happened?

_____

_____

_____

❹ When you added the flour, did the flour form a layer as the mixture settled? Why or why not?

_____

_____

_____

IV. Why?

In this experiment you observed the differences between rocks and dirt and how both rocks and dirt behave in a water mixture. When you mixed the rocks and dirt with water, you got what is called a slurry. When this slurry was allowed to settle, you noticed that the rocks settled first, and the lighter dirt particles settled last. When the flour was added, you could probably see a layer of flour settling between layers of rocks and dirt.

In your textbook you learned about sedimentary rocks. In this experiment you observed how sedimentary rocks are formed when sediments (rocks, dirt, and other particles) settle. The heavier particles will settle first, and the lighter particles will settle last. This causes layers to be formed.

Sedimentary rocks are formed when these layers are put under great pressure. When you see a sedimentary rock, you can observe different layers in the rock. These rock layers are created in the same way you created layers with your rock, dirt, flour, and water mixture.

V. Just For Fun

You can make your own edible sedimentary rock. Try the recipe below. Observe how the layers settle.

Get a cake mix and follow the directions on the package to make a batter. Before pouring the batter into a baking pan, stir in some nuts, gumdrops, and chocolate chips or M&Ms.

When you cut the cooled cake, observe the layers that have formed. In the box below, draw the cake and its layers.

Edible Sedimentary Rock

Experiment 4

The Shape of Earth

I. Think About It

Why do you think the Earth is shaped like a slightly smashed ball that is a little farther across than it is from top to bottom? Write or draw your ideas below.

_____
_____
_____
_____
_____
_____
_____

II. Observe It

❶ Take a baseball and place it on the floor. Try to spin the ball in place.

❷ Write or draw your observations below.

_____
_____
_____
_____
_____
_____
_____

❸ Fill a balloon with water, tie it closed, and place it on the floor.

❹ Without breaking the balloon, spin it in place. Write or draw your observations below.

_____
_____
_____
_____
_____
_____
_____

III. What Did You Discover?

❶ Did the baseball change shape when you spun it on the floor? Why or why not?

_____

_____

_____

❷ Did the water balloon change shape when you spun it on the floor? Why or why not?

_____

_____

_____

❸ Can you change the shape of the baseball with your hands? Why or why not?

_____

_____

_____

❹ Can you change the shape of the water balloon with your hands? Why or why not?

_____

_____

_____

IV. Why?

Since a baseball is solid all the way to its center, its shape cannot easily be changed. A water balloon has a soft center, which means that its shape can be more easily changed than can the shape of a baseball.

We live on the hard, rocky part of Earth (the crust), but Earth is mostly soft or fluid inside. The Earth spins around an axis, which is an imaginary straight line that goes through the center of the Earth. This spinning motion creates an outward-directed force called *centrifugal force*. Centrifugal force causes the Earth's center to bulge slightly as the soft part of the inner Earth is forced outward by the centrifugal force.

In this experiment the balloon that holds the water is similar to the crust and outer mantle of the Earth. Both the balloon and the hard outer layers of Earth surround a soft center. As the balloon is spun, it is able to change in shape in response to centrifugal force. The crust and outer mantle and the soft part of the inner Earth have also changed in shape due to the centrifugal force caused by the spinning of the Earth. This has created the bulge at the equator of the Earth.

(See *Focus On Elementary Astronomy—3rd Edition* for more about the effects of Earth's spin on its axis.)

V. Just For Fun

Sometimes a scientist will get an idea about what something is like or how it works. Then the scientist will do an experiment by thinking about that idea and all its possibilities. This is called a *thought experiment*.

Here is a thought experiment for you to try. Use your imagination while doing this thought experiment. Write or draw your ideas below and on the next page.

THOUGHT EXPERIMENT

We have discovered that the Earth has a slightly flattened ball shape. Think about what it might be like if, instead of being round, the Earth were shaped like a cube.

Do you think anything would be different if Earth were shaped like a cube? Would travel be different? What would happen when you came to an edge or a corner? What things can you think of that might be different?

Write or draw your ideas.

_____
_____
_____
_____
_____

Experiment 4: The Shape of Earth   43

_____
_____
_____

Thought Experiment
EARTH AS A CUBE

Experiment 5

Mud Volcanoes

In this experiment you will explore how the thickness of lava determines the type of mountain that can be formed by a volcano.

I. Think About It

❶ What happens when you pour thick syrup on your pancakes?

_____

_____

_____

❷ How fast does it flow from the pancakes to the plate?

_____

_____

_____

❸ What do you think would happen if you poured water on your pancakes?

_____

_____

_____

❹ How fast do you think the water would flow from the pancakes to the plate?

_____

_____

_____

❺ What happens when you mix dirt with a little bit of water? Does it make a good mud pie? Why or why not?

_____
_____
_____
_____
_____

❻ What happens when you mix dirt with a lot of water? Does it make a good mud pie? Why or why not?

_____
_____
_____
_____
_____

II. Observe It

Part I. Mix the following:

❶ Mix .5 liter (about 2 cups) of dirt with .25 liter (about 1 cup) of water and label it "**A**".

❷ Mix .5 liter (about 2 cups) of dirt with .5 liter (about 2 cups) of water and label it "**B**".

❸ Mix .5 liter (about 2 cups) of dirt with .75 liter (about 3 cups) of water and label it "**C**".

Part II. Observe your three mixtures. In the chart below, write the answers to the following questions.

① Is the mixture thick or thin?

② Can you form a mud pie with the mixture? Why or why not?

③ Which is the thickest mixture?

④ Which is the thinnest mixture?

Question	A	B	C
①			
②			
③			
④			

Part III. Now take each mixture (**A**, **B**, and **C**) and pour a little of each on the ground. Pour each mixture in its own area and label each area **A**, **B**, or **C**.

Observe what happens. In the chart below, write the answers to the following questions.

① Is the mixture easy or difficult to pour?

② Does the mixture stay together or spread out? Why?

③ How far does each mixture spread?

Question	A	B	C
①			
②			
③			

Part IV. After you pour the mixtures, allow the three areas to dry out. Then add another layer to each area. Pour the same mixture over the same spot where you poured it the first time. Add more layers, letting each layer dry out before adding another.

Observe what happens. In the chart below, write the answers to the following questions.

① Does the mixture form layers?

② How high are 2, 3, or 4 layers of each mixture?

③ How wide are 2, 3, or 4 layers of each mixture?

④ How far does each mixture spread?

Question	A	B	C
①			
②			
③			
④			

III. What Did You Discover?

❶ Which was the thickest mixture?

_____

❷ Which was the thinnest mixture?

_____

❸ How easy was it to pour the thickest mixture? Why?

_____

_____

❹ How easy was it to pour the thinnest mixture? Why?

_____

_____

❺ Which of the 3 mixtures would make the best cone volcano? Why?

_____

_____

_____

❻ Which of the 3 mixtures would make the best shield volcano? Why?

_____

_____

_____

IV. Why?

Volcanic lava begins as magma in the mantle of the Earth. When the magma finds its way to the surface, it is called lava. Lava is an extremely hot mixture of melted rocks and minerals.

Some lava is thick like mixture **A**, and some lava is thin like mixture **C**. When thick lava comes out of a volcano, it can form a cone volcano with steep sides. Because the lava is thick, it won't flow very far from the center. When thin lava comes out of a volcano, it can form a shield volcano. Because the lava is thin, it can flow very far away from the center of the volcano.

V. Just For Fun

❶ Using thick mud, build a cone-shaped volcano.

❷ Before the mud dries, use a pencil to poke a hole down the center of the cone.

❸ Allow the mud to dry.

❹ When it is dry, pour 15 milliliters (1 tablespoon) of baking powder down the hole.

❺ Next, gently pour in 15 milliliters (1 tablespoon) of vinegar.

❻ Observe what happens and draw a picture of it on the following page.

Mud Volcano

Experiment 6

All the Parts

Introduction

What if taking something apart can help you understand how it works?

I. Think About It

❶ What are the different parts of a bicycle?

___

❷ How does each part of a bicycle work?

___

❸ What are the different parts of a car?

___

❹ How do the different parts of a car work?

_____
_____
_____
_____
_____

❺ What are the different parts of an airplane?

_____
_____
_____
_____
_____

❻ How do the different parts of an airplane work?

_____
_____
_____
_____
_____

II. Observe It

❶ Find a toy or small piece of equipment that you can take apart (for example, a small music box or toy car). Before you disassemble the item, observe how it functions. Does it move? Does it make sound? Does it create light? Record your observations by writing and drawing.

❷ Using the appropriate tools and with the help of an adult, carefully take apart the toy, music box, small car, or other object. Disassemble it until you can no longer take anything apart. Record the number of parts you have.

_____

❸ Examine each part carefully. For each part record the shape, size, weight, and what you think it does. For example, if there is a gear, the gear may move something else. If there is a crank, the crank may rotate the gear, and so on. Use the following boxes to record your observations.

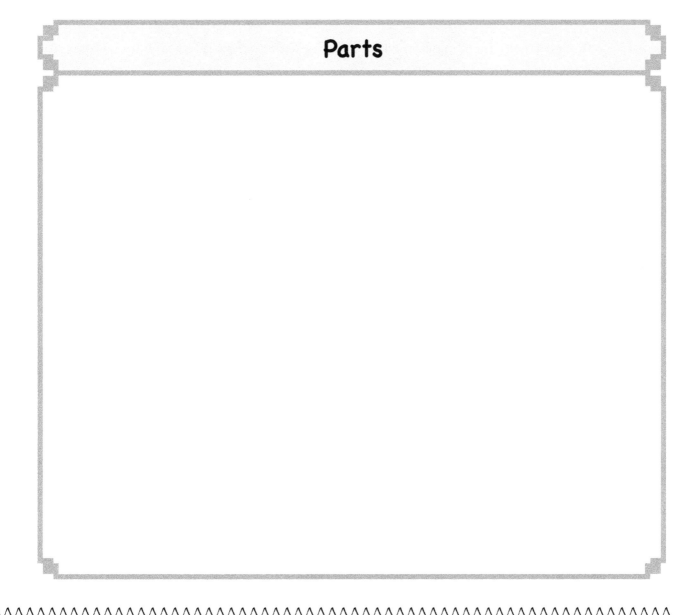

Parts

More Parts

❹ Reassemble the object you took apart and see if you can get it to work again. Record your observations.

Reassembly

III. What Did You Discover?

❶ How many parts did your item have?

_____

❷ How many different functions did you discover? What were they?

_____

_____

_____

❸ How easy or difficult was it to disassemble your item? Why?

_____

_____

_____

❹ How easy or difficult was it to reassemble your item? Why?

_____

_____

_____

❺ What did you learn about how your item works?

_____

_____

_____

_____

IV. Why?

The best way to learn about how something works is to simply take it apart. In the process of taking something apart, you can see where two parts are connected and what they do. You can also examine each part by itself in detail, learning how it is made, its size and shape, and how it fits together with other parts.

Scientists take things apart all the time. Scientists often spend their whole careers studying just one part of a whole system. In geology, scientists can spend years studying just one small part of the biosphere, or just the chemistry of the ocean, or just the water patterns in a river. By studying one part of the whole Earth in detail, geologists can learn a great deal about how the Earth works.

V. Just For Fun

Taking things apart is fun. Take apart another object and learn as much as you can about all the parts and how they work. Then reassemble the parts.

What object will you be taking apart?

_____

In the following boxes, record your observations about the object, its parts, and how the different parts work.

Observations

More Observations

Experiment 7

Edible Earth Parfait

Experiment 7: Edible Earth Parfait 65

Introduction

Scientists use many different materials to make models. Do you think you could make a model of Earth and then eat it?

I. Think About It

Think about Earth and its different layers. Now imagine that Earth's layers are made of sugar, salt, flour, gelatin, graham crackers, and ice cream. What if you could make Earth from some of your favorite food items?

❶ What food items could you use to make the hard crust of an edible Earth?

_____

❷ Would you use two different kinds of food items for an edible crust and an edible lithosphere? Why or why not?

_____
_____
_____

❸ What food items could you use to make the soft asthenosphere of an edible Earth?

_____

❹ What food items could you use to make an edible mesosphere?

_____

❺ What food items could you use to make an edible outer core?

_____

❻ What food items could you use to make an edible inner core?

_____

II. Observe It

❶ Decide which food items you will use to make an edible Earth. List each item and which of Earth's layers it represents.

_____

_____

_____

_____

_____

❷ Is your modeling of Earth limited by the food items that are available to you? Why or why not?

_____

_____

❸ Take a tall, clear glass and assemble the layers of your edible Earth in the same order they are found in the Earth.

❹ When you finish assembling your edible Earth, observe how well the layers remain separate. Note if any layers are moving down the sides and mixing with other layers. Also note if layers that are more solid behave differently than softer layers.

❺ Eat your Edible Earth Parfait before it melts!

❻ Look up the word *parfait*. Does your Edible Earth qualify as a parfait? Why or why not?

_____

_____

III. What Did You Discover?

❶ How many layers did you make? _____

❷ Why did you decide on this number of layers?

_____

_____

❸ Was it easy or difficult to assemble the layers? Why or why not?

_____

_____

_____

❹ Did any of the layers overlap or mix into one or another? Why or why not?

_____

_____

_____

❺ Did the layers change over time as you were assembling them? Why or why not?

_____

_____

_____

❻ Do you think doing this experiment helped you better understand Earth's layers? Why or why not?

_____

IV. Why?

Creating a model is a great way to explore the educated guesses about how Earth might be layered. Making a model is helpful in gaining an understanding of how Earth's layers could fit together and might change. Even though the food items you used to build an edible Earth are very different from the actual materials that make up the Earth, you may be able to observe some similar events.

For example, if you used a top crust made of graham crackers, an outer core of hot fudge, and a mesosphere and asthenosphere of ice cream, it's possible that some of the ice cream melted into the hot fudge. Geologists don't know if layers move into each other at Earth's core, but because layers in models behave this way, it can be seen as a good educated guess. Also, the ice cream in the middle may melt and push through small cracks in the graham cracker crust above. This is similar to what happens when lava pushes through the Earth's crust. By observing what happens in models, even edible models like an Earth parfait, scientists are able to make educated guesses about what happens with Earth's layers.

V. Just For Fun

Make an Inedible Earth Parfait! Review the questions in *I. Think About It* and decide what inedible items you could use for each layer. In the following box draw your Inedible Earth Parfait and label each layer with the materials you used.

Inedible Earth Parfait

Experiment 8

What's the Weather?

Introduction

This experiment is about exploring the atmosphere by observing the weather and its effects.

I. Think About It

❶ Do you think the sky looks different on different days? Why or why not?

_____

_____

❷ Do you think clouds always look the same? Why or why not?

_____

_____

❸ What do you think happens to plants when the wind blows hard?

_____

_____

❹ Do you think rain is important for plants? Why or why not?

_____

_____

❺ Do you think the weather changes the way animals behave? Why or why not?

_____

_____

II. Observe It

Observe the weather every day for a week. Is it sunny? Cloudy? What do the clouds look like? Is it hot or cold? Is it raining or windy? What happens to the ground, the plants, and the animals as the weather changes?

In the following boxes, write or draw your observations.

DAY 1
Temperature _____

Observations

DAY 2
Temperature _____

Observations

DAY 3
Temperature _____

Observations

Experiment 8: What's the Weather? 75

DAY 4
Temperature _____

Observations

DAY 5
Temperature _____

Observations

DAY 6
Temperature _____

Observations

DAY 7
Temperature _____

Observations

III. What Did You Discover?

❶ What changes in the weather were you able to observe?

_____

_____

_____

❷ Did changes in the weather cause any changes in animal behavior? If so, what were they?

_____

_____

_____

❸ Did changes in the weather affect the plants? If so, in what ways?

_____

_____

_____

❹ Was the ground affected by the weather? How?

_____

_____

_____

❺ In what ways did the weather affect your behavior and activities?

_____

_____

_____

IV. Why?

Weather patterns change from hour to hour, day to day, and season to season.

The temperature of the atmosphere changes frequently. Energy from the Sun heats the atmosphere during the day. At night some of the heat is released back into space, making the air cooler.

The amount of moisture in the air also causes weather changes. When the air is dry, the sky is clear. When the air contains more water vapor, there are clouds. If there is a lot of water vapor, it may rain or snow, bringing water to plants and animals so they can live.

Winds can blow dirt around and even break or uproot plants. Extremely strong winds, like those in tornadoes and hurricanes, can move whole buildings!

Changes in the seasons also affect the weather. Summer and winter are caused by the tilt of the Earth toward or away from the Sun.

All of these changes can affect how animals behave. Some animals may be out looking for food, or they may be staying in trees or in their burrows, waiting for the weather to change.

V. Just For Fun

Make a wind tester!

Take a helium-filled balloon outside and find a place to tie it. Observe the balloon during the day. Does the balloon tell you anything about the wind? Why or why not? Draw or write your observations below.

Wind Tester

Experiment 9

How Fast Is Water?

Introduction

Do you think anything different will happen if you pour water on sand or pour it on pebbles? Find out here!

I. Think About It

❶ When rain falls on the land, where do you think the rainwater goes?

_____
_____
_____
_____

❷ Does the kind of ground that rain falls on affect what happens to the rainwater? Why or why not?

_____
_____
_____
_____

❸ Do you think there is water under the ground? Why or why not?

_____
_____
_____
_____

❹ What do you think would happen to plants and animals if it did not rain? Why?

_____
_____
_____
_____

❺ Why do think rain makes mud puddles in some places but not others?

_____
_____
_____
_____

II. Observe It

❶ Take three Styrofoam cups and use a pencil to poke a hole in the bottom of each one.

❷ Measure about 240 milliliters (1 cup) of sand and pour it into one of the cups.

❸ Measure about 240 milliliters (1 cup) of pebbles and pour them into the second cup.

❹ Measure about 240 milliliters (1 cup) of small rocks and pour them into the third cup.

❺ Next, measure about 120 milliliters (4 ounces) of water and pour it into the cup containing the sand. Observe how long it takes the water to run through the sand.

❻ Write your observations in the box provided. In addition to length of time, record anything else you observe.

❼ Repeat Steps ❺ & ❻ with the pebbles and then with the small rocks. Compare how long it takes the water to run through each cup.

How Quickly or Slowly Water Runs	
SAND	
PEBBLES	
SMALL ROCKS	

III. What Did You Discover?

❶ Which material did the water run through the fastest? Why?

_____

_____

_____

❷ Which material did the water run through the slowest? Why?

_____

_____

_____

❸ Do you think all of the water came out of the cup with the sand? Why or why not?

_____

_____

_____

❹ Do you think all of the water came out of the cup with the small rocks? Why or why not?

_____

_____

_____

IV. Why?

In this experiment the water ran through the pebbles and the small rocks faster than it did through the sand. This happens because the particles of sand are much smaller than the pebbles and rocks. The small size of the sand particles allows them to fit together closely, leaving less empty space for the water to flow through. The pebbles and rocks have bigger spaces between the pieces, so water can flow through more quickly. The amount of empty space between the particles of a material is called its *porosity*.

The type of ground that rain falls on affects how much water can go into the ground. If rain falls on ground made of pebbles or small rocks, the rainwater will quickly flow through it. Rainwater also flows fairly quickly through sand and may not leave much water on the surface for animals to drink.

Another kind of soil is called clay. Clay has very small particles with little space between them. It can take water a long time to go through clay. Rainwater tends to form puddles on this type of soil. Clay can make very slippery and sticky mud!

The best type of soil for plants is in between sandy soil and clay soil and contains matter from decayed plants and animals. This kind of soil allows rainwater to flow through it slowly, keeping enough water for plants to use for making food.

V. Just For Fun

Mud City!

Make a mud city using dirt, water, pebbles, and rocks. Using what you learned in the previous sections of this experiment, make rivers and a lake that holds water.

You can use pebbles, rocks, sand, or dirt for city walls or to contain the rivers and lake. Think about how you can make the water stay in the lake and what will make water flow in rivers. Can you add things that will make the water flow faster or slower or change direction?

Once you've built your city, you can make paper boats to float on the rivers and the lake.

On the next page you can draw the mud city you created.

My Mud City

Experiment 10

What Do You See?

Introduction

If you explore the area where you live, do think you can observe things you haven't noticed before?

I. Think About It

❶ What living things do you think you will see if you walk around your yard and your neighborhood?

_____
_____
_____

❷ What do you think these living things eat?

_____
_____
_____

❸ Where do you think these living things sleep?

_____
_____
_____

❹ Do you think you will see different plants and animals in different areas? Why or why not?

_____
_____
_____

II. Observe It

❶ Go for a walk around your yard and your neighborhood. Take this *Laboratory Notebook* with you so you can draw and write your observations.

❷ Carefully observe the environment you're in by walking slowly and looking up to the sky, down to the ground, and everywhere in between. Look at things close up and from farther away.

❸ Look for different animals, bugs, birds, plants, and people. Observe what they are doing. Are birds flying in the sky, singing in trees, or hopping in the grass? Is a dog taking its person for a walk? Are there flowers blooming? Is a squirrel chattering at you from a tree? Are there bugs crawling on plants? Do you see any animals, birds, or bugs eating? What are they eating?

❹ Use the following boxes to draw and write what you see. There is a space in the gray boxes where you can note where you are when you make your observations.

Observations of the Environment

Observations of the Environment

Observations of the Environment

III. What Did You Discover?

❶ Did you observe anything about the plants in your environment that you hadn't noticed before? What did you observe?

_____
_____
_____

❷ What did you see animals doing?

_____
_____
_____

❸ Were any of the living things eating? What and how were they eating?

_____
_____
_____

❹ When you walked from one area of your environment to another, what differences in plants, animals, and birds did you observe?

_____
_____
_____

❺ What did you see that was surprising?

_____
_____

IV. Why?

The biosphere contains all the living things on Earth. Within the biosphere are different areas called *environments*. Each environment has a particular set of resources that make it possible for certain living things to live in that area. For example, a desert environment is a good home for lizards, cacti, and other animals and plants that don't need much water. An ocean environment is where you will find plants, fish, and animals, such as whales and dolphins, that can live in salty water. There are pine forest environments, mountain environments, lake and river environments, coastal environments, plains environments, and many others.

The region where you live has its own environment. The weather, amount of rainfall, type of soil, variations in temperature, plants, animals, and insects all work together to create this environment.

An environment can be any size. As you were making observations on your walk, you may have noticed that certain plants grow best in one area but not another, and certain animals live in one area but not another. This is a result of differences within the larger environment. Each plant and animal grows best when it finds just the right conditions to live in.

V. Just For Fun

An exoplanet is a planet that is orbiting a different sun (star) than ours.

Imagine you are the first astronaut to land on the exoplanet Kepler-62e in the constellation Lyra. Suppose you discover that Kepler-62e has a biosphere.

What do you think you would see when you land? Do you think Kepler-62e would look like Earth or be very different? Would there be an atmosphere with clouds? What would they look like? Would you find water? Where would the water be found? Would there be lots of plants and animals? What would the plants look like? What would the animals look like? What would they be doing? What would they eat? Would there be human-like creatures? What would they look like? What would they be doing? What else do you think you would see on Kepler-62e?

On the next page draw or write what you imagine the biosphere on Kepler-62e would look like.

A Visit to Kepler-62e

Experiment 11

Moving Iron

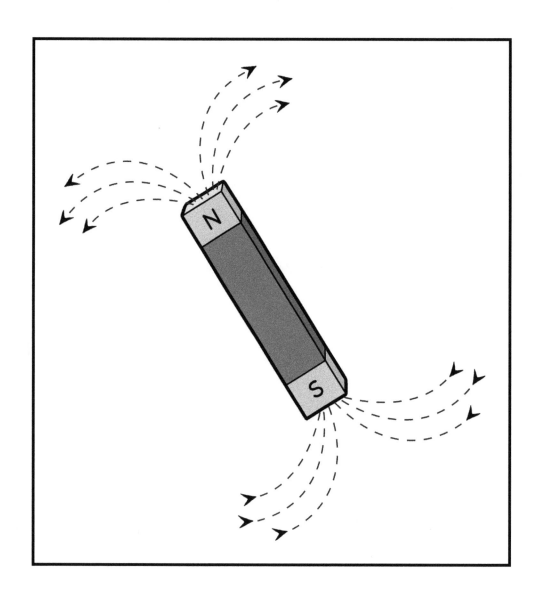

Introduction

Do you think there is a way to see the magnetic field created by a magnet? Find out with this experiment.

I. Think About It

❶ Do you think if you look at a magnet you will see its magnetic forces? Why or why not?

_____
_____
_____

❷ Do you think magnetic forces can be useful? Why or why not?

_____
_____
_____

❸ If you have two magnets, do you think you could make the two north poles stick together? Why or why not?

_____
_____
_____

❹ Do you think there is a way to show that Earth has a magnetic field surrounding it? Why or why not?

_____
_____
_____

II. Observe It

❶ Pour corn syrup into a plastic box until there is about 6 millimeters (1/4 inch) of syrup covering the bottom.

❷ Put the bar magnet on the table and place the box on top of it so the magnet is about in the center of the box.

❸ Pour iron filings on top of the syrup, being careful to not breathe them in.

❹ Wait 30 minutes and then check the iron filings.

❺ In the box below, draw what the iron filings look like.

III. What Did You Discover?

❶ What did the iron filings look like when you first put them in the syrup?

_____
_____
_____

❷ What did the iron filings look like after 30 minutes?

_____
_____
_____

❸ What do you think made the iron filings move?

_____
_____
_____

❹ Do you think if you did this experiment again, the iron filings would end up in the same pattern? Why or why not?

_____
_____
_____

❺ What do you think this experiment might tell you about Earth's magnetic field?

_____
_____
_____

IV. Why?

A magnetic field is the area around a magnet that is affected by magnetic forces. In this experiment you were able to "see" the magnetic field around your magnet by observing how it affected the iron filings.

Earth has a magnetic field similar to the one that affected the iron filings in this experiment. Earth's magnetic field points out from the North Pole, surrounds the Earth, and points in at the South Pole. Earth's magnetic field is different from that of a bar magnet because it is thought to be created by swirling molten metals in Earth's core rather than from the atoms in a solid bar of metal.

Another difference is that Earth's magnetic field extends into space and is affected by energy from the Sun. When the Sun's energy hits the magnetic field, it creates the magnetosphere. The magnetosphere, in turn, lets enough heat and light energy through to keep plants and animals healthy. At the same time, it stops too much energy from getting to Earth. Life could not exist if too much of the Sun's energy reached the surface of Earth, so the magnetosphere is essential for life to exist.

V. Just For Fun

❶ Try moving the box so the magnet is in a different position under it. Wait about 30 minutes, then look at the pattern the iron filings make. Did repositioning the magnet make a difference in the pattern?

In the box below draw the pattern made by the iron filings.

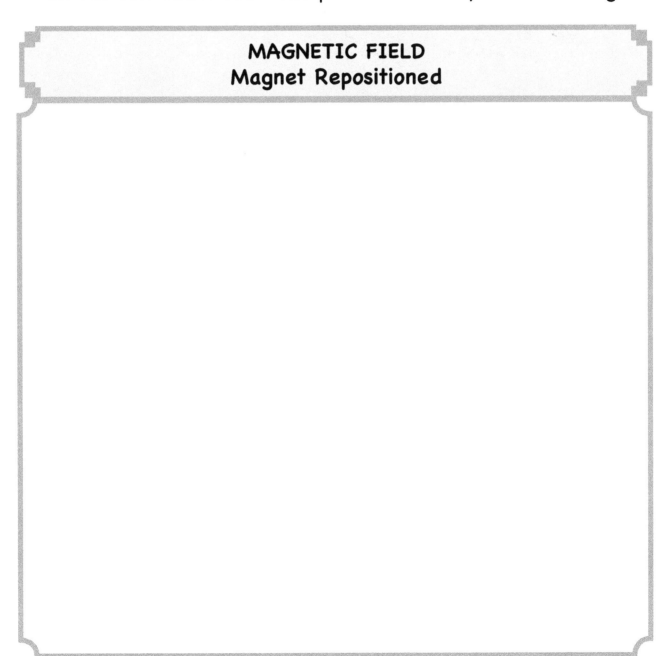

**MAGNETIC FIELD
Magnet Repositioned**

❷ Place a second magnet under the box. Wait about 30 minutes. Does having two magnets change the pattern of the magnetic field?

Draw your observations below.

**MAGNETIC FIELD
2 Magnets**

❸ Try moving the two magnets to different positions. Wait about 30 minutes each time, then check to see what the pattern of the iron filings looks like.

Draw your observations below.

**MAGNETIC FIELD
2 Magnets Repositioned**

Experiment 12

What Do You Need?

Introduction

Do you think you can observe how the different spheres of Earth work together? Try this experiment.

I. Think About It

❶ Do you think a plant needs the atmosphere in order to live and grow? Why or why not?

_____

_____

❷ Do you think a plant needs the hydrosphere in order to live and grow? Why or why not?

_____

_____

❸ Do you think a plant needs the biosphere in order to live and grow? Why or why not?

_____

_____

❹ Do you think a plant needs the geosphere in order to live and grow? Why or why not?

_____

_____

II. Observe It

Plant a garden!

❶ Decide what kind of plant you want to grow and then get seeds for it.

❷ Plant several seeds outdoors in a garden. If you don't have space for a garden, you can use a flower pot, milk carton, or other container with soil in it.

❸ Water the seeds when you plant them and then check the soil daily. When it starts to get dry, add more water.

❹ Write or draw your observations in the following boxes. Each box can have observations for more than one day.

❺ As your plant gets bigger, what do you observe about its growth? How fast is it growing? Is it staying healthy? Is it getting too much or too little water or sunshine? Are bugs eating the plant? If so, what do they look like? Do you see any worms in the soil? Are any animals or birds affecting your plant? What else can you observe?

❻ Write or draw your observations as the plant grows. As you make observations about your plant, think about which of Earth's parts is affecting it and in what way. Is it the biosphere, hydrosphere, atmosphere, geosphere, or magnetosphere, or is it a combination of two or more of these?

GROWING A PLANT

GROWING A PLANT

GROWING A PLANT

GROWING A PLANT

III. What Did You Discover?

❶ How easy or difficult was it to grow your plant? Why?

_____
_____
_____

❷ Did your plant get enough water from rain? How could you tell?

_____
_____
_____

❸ Did your plant have any problems with bugs? Why or why not?

_____
_____
_____

❹ Were any animals helpful or harmful to your plant? Why or why not?

_____
_____
_____

❺ Was there enough sunlight for your plant? How can you tell?

_____
_____
_____

❻ Was the soil you used good for your plant to grow in? Why or why not?

_____
_____
_____

❼ In what ways do you think the different parts of the Earth worked together to help your plant grow?

_____
_____
_____
_____
_____
_____
_____

IV. Why?

For a plant to grow, all the different parts of the Earth have to work together. The plant is part of the biosphere and so are the bugs, animals, and people that might eat the plant. Bacteria in the soil fix nitrogen for plants to use, and worms add nutrients to the soil.

Earth's geosphere provides soil that has minerals the plant needs for making food. The soil contains water that is taken up by the plant's roots. The soil also provides a place for roots to anchor the plant to the ground so it won't blow away in the wind.

The hydrosphere provides the water needed by the plant, whether it comes from rain or from your garden hose. The hydrosphere works with the atmosphere to make clouds that move over the Earth, carrying rain to different parts of the land.

The atmosphere has carbon dioxide for the plant to use to make food. The gases in the atmosphere hold onto some of the heat energy from the Sun, helping to keep Earth warm at night. Also, the atmosphere lets light energy from the Sun reach Earth so plants can use the Sun's energy for making food.

The magnetosphere protects the plant from getting too much energy from the Sun, which would be harmful.

If any one of these parts of Earth were missing, plants could not live and grow.

V. Just For Fun

Grow an herb garden!

Herbs are plants that are great for adding extra flavor to salads, soups, and other foods. Sometimes they are used to make delicious teas or for medicines.

Herbs can be grown outdoors in pots or in a garden bed, and many can be grown in small pots indoors.

Decide which herbs you would like to grow. You might look for information about herbs in the library or online. Or you might want to choose them by names that sound interesting or by what the plants in the store look like. Some possibilities are: basil, rosemary, thyme, chervil, lemon balm, oregano, parsley, dill, spearmint, chamomile. If you have a cat, it might enjoy having some catnip.

Once you have decided which herbs you want to grow, get seeds or small plants.

Tend your herb garden, and when the plants are big enough, gather some leaves to put in your salad and soup. While you and your family are eating the tasty herbs, you can tell everyone how all the parts of the Earth worked together to bring each plant's unique flavor to you—Earth's geosphere, biosphere, hydrosphere, atmosphere, and magnetosphere all helped out!

More REAL SCIENCE-4-KIDS Books
by Rebecca W. Keller, PhD

Building Blocks Series yearlong study program — each Student Textbook has accompanying Laboratory Notebook, Teacher's Manual, Lesson Plan, Study Notebook, Quizzes, and Graphics Package

Exploring the Building Blocks of Science Book K (Activity Book)
Exploring the Building Blocks of Science Book 1
Exploring the Building Blocks of Science Book 2
Exploring the Building Blocks of Science Book 3
Exploring the Building Blocks of Science Book 4
Exploring the Building Blocks of Science Book 5
Exploring the Building Blocks of Science Book 6
Exploring the Building Blocks of Science Book 7
Exploring the Building Blocks of Science Book 8

Focus Series unit study program — each title has a Student Textbook with accompanying Laboratory Notebook, Teacher's Manual, Lesson Plan, Study Notebook, Quizzes, and Graphics Package

Focus On Elementary Chemistry
Focus On Elementary Biology
Focus On Elementary Physics
Focus On Elementary Geology
Focus On Elementary Astronomy

Focus On Middle School Chemistry
Focus On Middle School Biology
Focus On Middle School Physics
Focus On Middle School Geology
Focus On Middle School Astronomy

Focus On High School Chemistry

Super Simple Science Experiments

21 Super Simple Chemistry Experiments
21 Super Simple Biology Experiments
21 Super Simple Physics Experiments
21 Super Simple Geology Experiments
21 Super Simple Astronomy Experiments
101 Super Simple Science Experiments

Note: A few titles may still be in production.

Gravitas Publications Inc.
www.gravitaspublications.com
www.realscience4kids.com

CPSIA information can be obtained
at www.ICGtesting.com
Printed in the USA
LVHW060150210819
628392LV00016BA/658/P